Heinemann InfoSearch

Sweeping Tsunamis

Heinemann Library
Chicago, Illinois

Louise and Richard Spilsbury

Customer Service 888-454-2279
Visit our website at www.heinemannlibrary.com

Designed by David Poole and Paul Myerscough
Originated by Dot Gradations Limited
Printed in China by W K T

07 06 05 04
10 9 8 7 6 5 4 3 2

Library of Congress Cataloging-in-Publication Data
Spilsbury, Louise.
  Sweeping tsunamis / Louise and Richard Spilsbury.
      v. cm. -- (Awesome forces of nature)
Includes bibliographical references and index.
Contents: What is a tsunami? -- What causes a tsunami? -- Where do tsunamis happen? -- What happens in a tsunami? -- Case study: Hawaii, 1946 -- Who helps when tsunamis happen? -- Case study: Japan, 1993 -- Can tsunamis be predicted? -- Can people prepare for tsunamis? -- Case study: Papua New Guinea, 1998 -- Can tsunamis be prevented? -- Destructive tsunamis of the past.
  ISBN 1-4034-3725-4 (lib. bdg.) -- ISBN 1-4034-4233-9 (pbk.)
  1. Tsunamis--Juvenile literature. 2. Tsunamis--Environmental aspects--Juvenile literature. [1. Tsunamis.] I. Spilsbury, Richard, 1963- II. Title.
  GC221.5. S69 2003
  363.34'9--dc21
                          2003001200

Acknowledgments
The author and publisher are grateful to the following for permission to reproduce copyright material:
Cover photograph by AGE Fotostock/Imagestate.
Unlike in this picture, tsunamis rarely break as they reach the shore.

pp. 4, 5, 9 Pacific Tsunamis Museum/ NOAA; p. 7 Charles O'Rear/Corbis; pp. 11, 14, 17 Pacific Tsunamis Museum; p. 12 Albert Yamauchi/Honolulu Star-Bulletin; p. 13 Mary Plage/Oxford Scientific Films; p. 15 Todd A. Gipstein/Corbis; p. 18 Scanpix Nordfoto/EPA/AFP; p. 19 Associated Press; p. 20 Bettmann/Corbis; p. 21 Rex Features; p. 22 David Butow/Corbis; p. 23 Carlos Munoz-Yague/LDG/EURELIOS/Science Photo Library; p. 24 Te Papa Tongarewa Museum of New Zealand; p. 25 Science Photo Library; p. 26 NOAA; p. 27 Wolfgang Kaehler/Corbis; p. 28 Getty Images.

Some words are shown in bold, **like this.** You can find out what they mean by looking in the glossary.

Contents

What Is a Tsunami?

A tsunami is a huge destructive ocean wave. It is nothing like an ordinary wave. As ocean waves move into shallow water, their narrow foaming tips curl over and "break," or collapse. A tsunami hits land as a dark, fast-moving ledge of water that rarely breaks as it nears shore. Most tsunamis are barely noticeable in deep parts of oceans, but they get bigger as they approach land.

TSUNAMI ⚡ FACTS

! The biggest tsunamis are the most destructive waves on the planet.

! The fastest tsunamis in the world can reach speeds of 500 miles (800 km) per hour.

! Tsunamis have reached heights of 130 feet (40 meters) above the normal level of the sea.

There are very few photographs of tsunamis because people do not usually stand around long enough to take them! This is because big tsunamis may move toward land at hundreds of miles per hour. This tsunami was photographed in Hilo, Hawaii, in 1946.

Awesome force

Big tsunamis are like huge walls of water. They can be tens of feet tall and a few miles wide, containing millions of tons of water. The water smacks hard onto land with the same force as a wall of concrete.

Anything in the path of a big tsunami—from people to giant ships or trucks—may be swept away, crushed, or buried under water. Trees and telephone poles are snapped like matchsticks. Homes, schools, and lighthouses may collapse as if made of cardboard. Over the past 100 years, tsunamis have killed tens of thousands of people and caused millions of dollars' worth of damage around the world.

Harbor waves

Tsunami is a Japanese word that means "harbor wave." It was given this name because of the great devastation caused around the coastal harbors of Japan by many tsunamis.

Smaller tsunamis may come ashore like a quickly rising tide, gently flooding coastal land. This flooding on Midway Island in 1952 was caused by a tsunami.

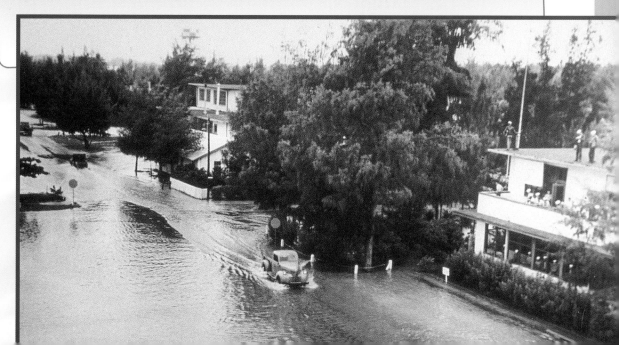

What Causes a Tsunami?

Tsunamis usually happen when giant chunks of land at the bottom of the ocean drop down as the result of an **earthquake.** Millions of tons of seawater move in to fill the gap. This water movement causes a series of waves on the surface of the ocean—a bit like the ripples that spread out when you drop a stone into a pond or lake.

Earth movements

The outer layer of Earth is made of solid rock. On mountain tops it may be bare, on deserts it may be covered with sand, and in oceans it is covered with seawater. Incredibly, this rock is always moving, although it does so very slowly. Deep inside Earth it is so hot that the rock is melted into a sticky liquid. The cooler, lighter rock of the surface floats around on top of this liquid in enormous chunks called **plates.**

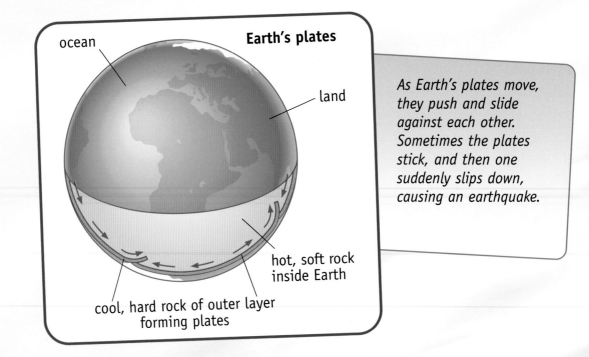

ocean

Earth's plates

land

hot, soft rock inside Earth

cool, hard rock of outer layer forming plates

As Earth's plates move, they push and slide against each other. Sometimes the plates stick, and then one suddenly slips down, causing an earthquake.

Other causes of tsunamis

All tsunamis start when massive amounts of seawater are suddenly moved. Sometimes the **lava** in Earth spurts out through gaps or thin spots in the plates. This is what we call a **volcano.** When underwater volcanoes explode, they destroy rocks around them and can start tsunamis.

Tsunamis can also be started when large amounts of rock or ice on mountains suddenly break free and fall into water. Tsunamis would also happen if a large meteorite—a piece of rock from space—plunged into an ocean.

In 1883, the Krakatoa volcano in Indonesia erupted. The whole island collapsed and caused 115-foot (35-meter) high tsunamis that sped toward neighboring islands, killing 36,000 people.

Deep beginnings

Tsunamis move outward from the point where they start. Imagine you are on a plane flying over an ocean. If an **earthquake** struck hundreds of feet under water on the **seafloor,** all you would see is crumpled water for an instant. The sea would then flatten again and a tsunami would speed away.

Tsunamis travel fastest in deep water. The **crest** of a tsunami in deep water may be only three feet (one meter) tall. This crest is just the tip of a deep wave that reaches tens of feet into the water. As the wave moves toward shallower water, the top pushes forward at a fast speed, but the bottom slows down as it touches the seafloor. The water then bunches up and the tsunami reaches its tallest height as it arrives at the coast.

Tsunamis can build to great heights as they get closer to land. Although they slow down, they can still hit coastlines at hundreds of miles per hour.

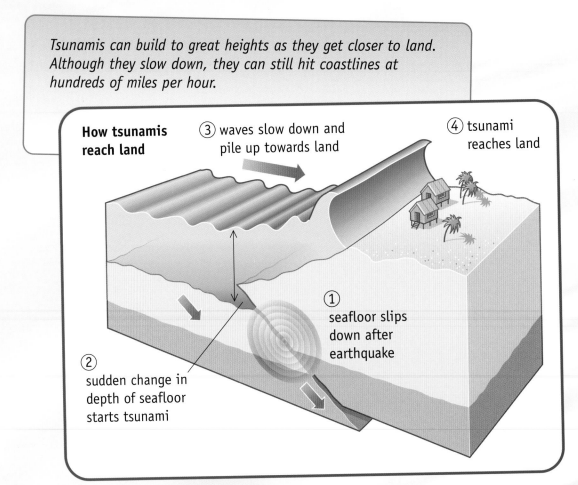

How tsunamis reach land

③ waves slow down and pile up towards land

④ tsunami reaches land

① seafloor slips down after earthquake

② sudden change in depth of seafloor starts tsunami

Enormous forces

Tsunamis shift huge quantities of water at such high speeds that they can travel very long distances. In 1960, an earthquake off Chile in South America started a tsunami. The tsunami traveled more than 9,000 miles (15,000 km) in 22 hours before hitting the coast of Japan. A big tsunami is barely slowed down when it moves over a small island, but it usually stops after it hits a **continent.** Some large tsunamis bounce off continents and move back and forth over whole oceans, getting gradually weaker over several days.

Tsunamis and tidal waves

Tsunamis are sometimes incorrectly called tidal waves. Tidal waves are waves caused by tides. Tides are the regular rise and fall of the level of the oceans, caused by the pull of **gravity** from the Moon and the Sun. Especially high tides sometimes cause large tidal waves, but never tsunamis.

Just a few inches of moving water can knock over a standing person. Imagine the damage several feet of water, like this tsunami in Hilo, Hawaii, in 1946, can do.

Where Do Tsunamis Happen?

Most tsunamis happen in the Pacific Ocean. Some of the countries most at risk from tsunamis are Japan, the United States, Papua New Guinea, and Chile, because they border the Pacific. Tsunamis happen in the Pacific Ocean because a part of Earth's outer surface, called the Pacific **plate,** lies underneath it. There are many **earthquakes** and **volcanoes** along the edges of this plate, where it meets other plates. This area is often called the "ring of fire."

Where else do they happen?

Tsunamis affect other coasts where earthquakes happen in the ocean. They have hit the east coast of Canada, which is at the edge of the Atlantic Ocean, and Turkey and Greece, which are in the southern Mediterranean Sea.

The countries around the edge of the Pacific Ocean are all at risk from tsunamis. Certain Pacific islands, such as Hawaii, are at even greater risk because they are in the middle of the ring of fire. Tsunamis can approach from all sides!

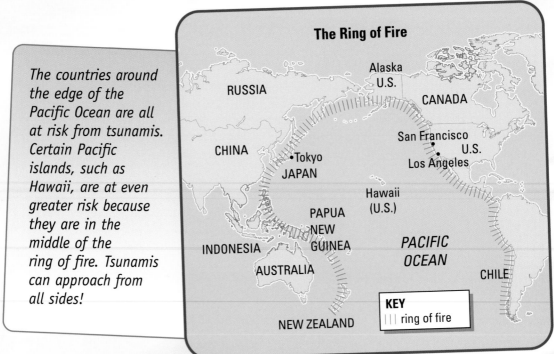

The Ring of Fire

RUSSIA

Alaska
U.S.

CANADA

San Francisco
• U.S.
Los Angeles

CHINA

•Tokyo
JAPAN

Hawaii
(U.S.)

PAPUA
NEW
GUINEA

PACIFIC
OCEAN

INDONESIA

AUSTRALIA

CHILE

KEY
||| ring of fire

NEW ZEALAND

Shape of the land

Some parts of coasts are more affected by tsunamis than others. Towns and villages in greatest danger are those at **sea level** less than 1.25 miles (2 km) from the sea. Even very small tsunamis can travel a long way over flat land.

Tsunamis are also dangerous in curved bays or at the end of **fjords.** The waves get very high between their narrow sides. When tsunamis reach a **headland**—a narrow strip of land sticking out into the sea— they wrap around it. Then water floods onto land from both sides.

TSUNAMI FACTS

! The highest tsunami ever recorded happened in 1958 in Lituya Bay, Alaska. A **landslide** fell into the narrow fjord causing a wave more than 1,600 feet (500 meters) high—that's taller than the Sears Tower!

Tsunamis can roll much farther inland over a flat coastline like this, than they can over a steep or hilly shore.

What Happens in a Tsunami?

Tsunamis move very fast. If someone sees one approaching, it is probably too late for him or her to get away from it. Sometimes, though, there are signs that a tsunami is on its way.

Many tsunami survivors describe how the **sea level** drops. Water is suddenly sucked away from the shore, uncovering sand, mud, and reefs on the **seafloor** and leaving fish and boats stranded. The water has moved to fill the space on the ocean floor created by an **earthquake.** Then, the water returns in waves.

"It looked as if someone had pulled the plug out from the seabed." C. Tayfur, a survivor of a tsunami that hit Turkey in 1999.

*Imagine how interesting it is to see a **coral reef** exposed and fish stranded on the shore. Sadly, many people attracted to a sight like this have been the first to feel the effects of a tsunami when it hits.*

Wave train

As the sea gurgles out from land, there is sometimes a very strong wind. This air is being pushed in front of the speeding tsunami. A big tsunami often comes in a series of waves called a wave train. The time between each wave **crest** may be minutes or even as long as an hour. Between each tsunami crest there is a **trough,** when water is again sucked out to sea. It is as if the water is being pulled by an enormous vacuum cleaner before it rushes back.

The first tsunami wave may not be the worst—the biggest, most dangerous waves in a wave train are often the third and eighth waves to arrive. After a tsunami strikes, it may take days before ocean waves get back to their regular sizes.

The first tsunami wave to break may not cause the most damage. Other, more damaging waves, may arrive later.

Destruction

A big tsunami can destroy nearly anything in its path. In an instant, whole areas of homes, farms, and factories may be ruined. Many animals and people may be drowned under several feet of water, or they might be carried as far as half a mile inland. Cars, trains, boats, buses, and shattered buildings and bridges are carried inland at high speed, like missiles. They may crush anything in their way.

An entire coastline may be altered by a tsunami. The seawater may flood large areas of low-lying land, ruining farmers' **crops.** Trees, other plants, and soil are sometimes stripped from the land, and the sucking action of the wave train may shift whole beaches.

This boat has been carried inland by a tsunami. A tsunami is incredibly strong and can move at the speed of a jet plane.

After the waves

When a tsunami is over, daily life does not return to normal for some time. Many people, such as fishermen with broken boats, cannot earn money because they cannot work. Children may not be able to go to school. There are many health hazards. Drinking water and **sewage** get mixed with seawater when pipes are snapped and **reservoirs** broken. People might drink **polluted** water containing germs that will make them sick. Many people are at risk of **electrocution** from damaged **powerlines.** Sometimes gas that leaks from broken pipes explodes.

When the seawater drains away, massive amounts of **debris** left on land have to be cleared away. Some debris comes from collapsed buildings or trees. Other debris is from the ocean—tsunamis pick up tons of sand, coral, rock, and fish as they approach land.

After a tsunami, there is often a terrible odor from piles of rotting fish that the tsunami dumped on land.

Hawaii, 1946

At around midday on April 1, 1946, an **earthquake** shook the Aleutian Islands of Alaska. A tsunami sped away across the Pacific Ocean. Five hours later, people in the town of Hilo on the eastern side of Hawaii were celebrating April Fools' Day. They were enjoying themselves near the harbor when, suddenly, the sea pulled back hundreds of feet. Some people walked down onto the seabed to see the **coral reefs** and gasping fish. The 26-foot (8-meter) high **crest** of the first wave caught most people by surprise.

> "The wave flipped me over and carried me toward the **lava** rock wall that rimmed the school. I recall telling myself, 'I'm going to hit head first into that rock wall and . . . die.' Miraculously, part of the wave that preceded me smashed into the wall and broke it up. So I went flying through the wall . . . rolling with all the rocks." M. Kino, a survivor of the 1946 tsunami

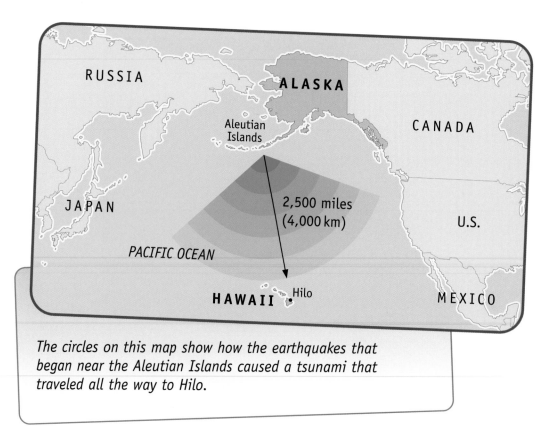

The circles on this map show how the earthquakes that began near the Aleutian Islands caused a tsunami that traveled all the way to Hilo.

Damage

About 100 people in Hilo were killed by the tsunami and hundreds more were injured. More than 1,000 buildings were damaged and the costs of **aid,** cleaning up, and rebuilding were around $25 million.

Most people in Hilo had no warning that a tsunami was coming. After Hilo, the U.S. government decided to create a tsunami early warning system around the Pacific Ocean. When another big tsunami approached in 1960, warning sirens were sounded and fewer people were killed or injured. Sadly, many of the victims had ignored the warnings and had not moved to a safer place.

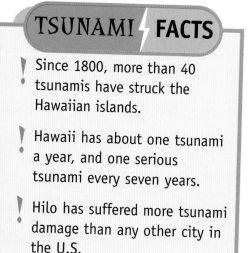

TSUNAMI FACTS

- Since 1800, more than 40 tsunamis have struck the Hawaiian islands.

- Hawaii has about one tsunami a year, and one serious tsunami every seven years.

- Hilo has suffered more tsunami damage than any other city in the U.S.

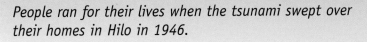

People ran for their lives when the tsunami swept over their homes in Hilo in 1946.

Who Helps when Tsunamis Happen?

Many people help the victims of tsunamis. Scientists tell coast guards and government officials if there has been an **earthquake** that might start a tsunami. Then fire, police, ambulance services, and the navy are put on alert. Hospitals prepare to treat tsunami victims.

Authorities use radio and TV **broadcasts** to warn people that they may have to **evacuate.** Authorities might also visit people where they live to warn them. If a tsunami is definitely on its way, sirens may be sounded so that people know they must evacuate to a safe place. Most people can travel on their own, but some young, old, or sick people need help evacuating. The police and army usually help organize evacuations.

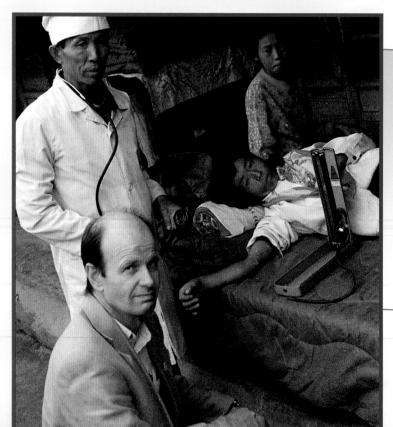

*Local authorities and **charities,** such as the Red Cross, give **first aid** to injured people. They also collect food and water supplies and organize shelter for evacuated people.*

Emergency help

More help is needed if a tsunami happens unexpectedly or if it affects more land than expected. Emergency services find and rescue people who were washed out to sea or trapped in dangerous places such as boats or unsafe buildings. The fire department puts out fires of spilled oil or gas. **Paramedics** and doctors give first aid for injuries such as cuts and broken bones.

In some poorer countries, emergency services may not be able to cope with a tsunami disaster. They will ask for **aid** from other governments and international organizations and charities. Many of these groups may help over a long period, working with local people to rebuild homes and hospitals. In places where farms have been destroyed, they give seeds and **livestock** so people will be able to feed themselves again. They also give new pipes and pumps to provide a safe water supply.

This rescue team has landed in a helicopter. They have search and rescue dogs to help them find people who may be trapped in collapsed buildings.

What to do when a tsunami happens

- People who live or work on the coast should move at least 1 mile (1.6 km) inland or to higher ground.
- People on a boat at deep sea should stay there and not return to land. The farther out to sea over deep water, the lower the wave height.
- People should keep a radio or telephone nearby to listen for further warnings and messages.
- People should never return to a tsunami-hit area until they are told the waves have stopped coming.
- People should stay away from buildings left standing after a tsunami because they might collapse.

People on boats near coasts should head out to sea if they hear a tsunami warning. Boats near the coast will be tossed and smashed onto the shore like toys, like this boat in Hachinohe, Japan, in 1960.

Japan, 1993

On July 12, 1993, an **earthquake** struck 12.5 miles (20 km) off the island of Okushiri. The Japanese authorities gave a tsunami warning on radio and TV within five minutes, so many people were able to **evacuate** to higher ground. However, waves 16 to 100 feet (5 to 30 meters) high had already struck Aonae, a fishing village on Okushiri's southern **headland.** More than 200 people were killed by the waves.

The Japan Maritime Safety Agency used helicopters, boats, and divers to find missing people. Heavy cranes and bulldozers were used to clear **debris** that had filled the harbor.

The gymnasium of Aonae Middle School became a temporary shelter for hundreds of people whose homes were destroyed. Their new homes were built farther away from the sea, so any future tsunamis would affect them less.

This photograph of Aonae was taken the day after the tsunami in 1993.

Aonae

SEA OF JAPAN

JAPAN

PACIFIC OCEAN

Can Tsunamis Be Predicted?

Scientists around the world work together to predict tsunamis. They use machines called **seismographs** to record when and where **earthquakes** strike and how strong they are. If a big earthquake happens under or near an ocean, scientists warn neighboring countries that there may be a tsunami. However, not every earthquake starts a tsunami, so scientists have developed ways of detecting and tracking tsunamis.

Detecting waves

There is a network of research stations around the Pacific Ocean that measure the heights of tides. Scientists use these stations to notice sudden changes in water level that happen before a tsunami. Scientists from Japan and the United States have laid cables on the ocean floor, near their coasts. Special boxes along the cables sense when a deep tsunami wave passes over them. The boxes send this information via **satellites** to scientists on shore so they know a tsunami is on its way.

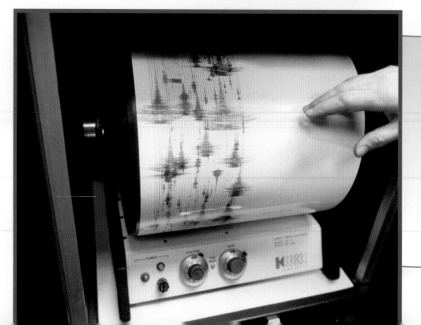

The marks on this paper show earth movements recorded by a seismograph. The more the ground shakes, the bigger an earthquake is.

Looking into the future

Many countries use powerful computers to help them **simulate** how tsunamis will affect their coasts. To do this, they prepare electronic maps of coastal places, showing where people live and what the land is like—for example, if it is flat or hilly. They also enter information about earlier tsunamis, such as how far they traveled on flat land and up steeper slopes. Then, they simulate tsunamis of different sizes and speeds hitting these places.

These simulations are vital to figure out the best places to **evacuate** people. Local authorities can then tell people the quickest escape routes if a tsunami approaches. Simulations also show the safest places to build.

These are computer simulations of tsunamis hitting land. They give people an idea of what might happen if a tsunami were to hit their coast.

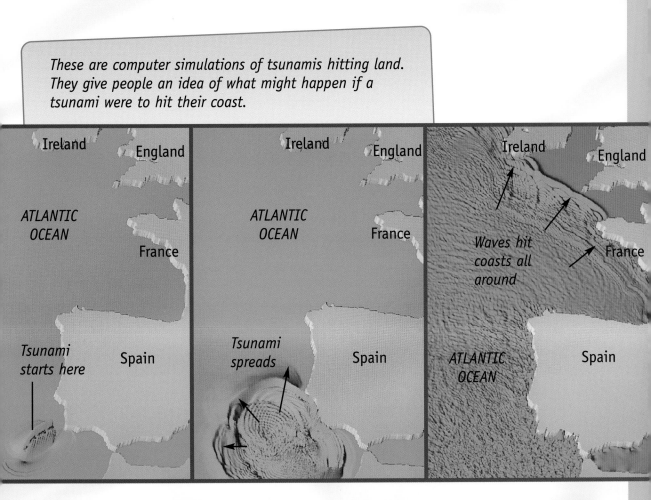

Can People Prepare for Tsunamis?

People who live in places threatened by tsunamis must understand the danger they are in. They can then prepare in various ways, from the way they build their houses to knowing how to **evacuate.**

Building for tsunamis

The damage caused by tsunamis can be reduced if buildings are made stronger. For example, most of the repair costs after the Aonae tsunami were spent fixing damaged harbor walls. Since then, new taller and stronger sea walls have been built around parts of Japan. In Hawaii, many office and hotel buildings are now built on stilts—the ground floor is open parking space with rooms above. Water can pass through the open space and not damage the building structure.

This museum in New Zealand is shaped so that waves can move around it. It is also built on rubber feet so waves can move it 2.5 feet (75 cm) without damaging it.

Evacuation plans

It is important for people to know the best way of getting themselves and their family, pets, and farm animals to a safe place. All family members should know how to contact each other or where to meet if they are not all in the same place. They should also know things about where they live and understand evacuation warnings. People should always take evacuation warnings seriously. They should also watch for tsunami signs such as sudden changes in **sea level.**

A tsunami survival kit

If people have to evacuate, they should take these things:

- a **first-aid** kit and any necessary medicines;
- enough food and water for at least three days;
- a flashlight and radio with extra batteries;
- warm clothing, blankets, or sleeping bags;
- money and important papers such as passports and driver's licenses.

States bordering the Pacific Ocean have put up tsunami signs like this one. They show people which way to go if they hear a tsunami warning on the radio or television.

Papua New Guinea, 1998

In July 1998, an **earthquake** started a giant underwater **landslide** that caused a huge tsunami. It was dusk when three massive waves swept over Arop and other villages around Sissano lagoon in northern Papua New Guinea.

Earthquakes are common in Papua New Guinea because it is near the joint between two of Earth's **plates.** However, most of the local people did not know about tsunamis. A rumble was heard out at sea and many people moved closer to see the water drawing back and then rising above the horizon. Those who could, ran for their lives. The first wave flooded the land and broke up the villagers' flimsy wooden houses. The second wave, which was 32 feet (10 meters) high, swept away everything in front of it. Two whole villages were completely destroyed. Thousands of people were injured or killed, many by being thrown against trees or being hit by floating **debris.** Many of the victims were children.

Before

After

These pictures show a village center before and after the tsunami of 1998. Whole buildings were wiped out by the tsunami's awesome power.

Sissano

PAPUA NEW GUINEA

PACIFIC OCEAN

Future safety

After the disaster, Australian emergency services and **charities,** such as Oxfam, worked with the government of Papua New Guinea to take care of survivors. They provided **aid** and helped them rebuild their lives. They also showed local people what to do if another tsunami happened.

Charity workers and the government of Papua New Guinea produced posters, leaflets, special TV programs, and videos about tsunamis. They visited all of the coastal villages and told local people what to do if a tsunami happens. In November 2000, another, smaller tsunami struck. Amazingly, no one was killed. Most people had remembered their lessons. They had climbed to high ground as soon as the earth began to shake and the **sea level** changed.

In Papua New Guinea, children are now taught all about tsunamis at school. Teachers explain that Papua New Guinea has been hit by many earthquakes and tsunamis, and they tell children what to do if a tsunami happens.

Can Tsunamis Be Prevented?

Tsunamis are natural events that have happened throughout Earth's history. Just as we cannot stop **earthquakes** or **volcanoes,** we cannot prevent tsunamis. There has been at least one tsunami a year in the Pacific Ocean since 1800, and there will be more in the future. Big tsunamis that affect the coasts of many countries happen about once every seven to ten years.

Changing world

As the population of the world grows, coastal towns and cities get bigger. This growth means that each tsunami may affect more people. Some states on the Pacific coast have large areas of sand and gravel. It was dropped there by big tsunamis in the past, when few people lived there. If a big tsunami happened now, millions of people on the West Coast might be affected by tsunami **debris.**

The key to dealing with tsunamis in the future is to be prepared. Learn what they are and what to do if one happens.

Destructive Tsunamis of the 20th Century

1908, Calabria, Italy
After an earthquake, a 50-foot (15-meter) high tsunami killed almost 100,000 people.

1933, Sanriku, Japan
An earthquake created a tsunami that killed 3,000 people, destroyed 9,000 buildings, and overturned 8,000 boats. The waves traveled as far away as Iquique, Chile.

1958, Lituya Bay, Alaska
An earthquake shook 90 million tons of icy rock from a **fjord** wall. It caused an enormous tsunami that carried anchored boats out to sea and stripped the fjord walls of all their soil and plants.

1960, Hilo, Hawaii
An earthquake in Chile sent 30-foot (10-meter) waves toward many Pacific islands. Waves killed 100 people in Japan. In Hilo, Hawaii, more than 200 buildings were destroyed.

1964, Anchorage, Alaska
A tsunami sped along the Pacific coast at more than 370 miles (600 km) an hour, crushing houses in its path. Later, it hit Crescent City, California, destroying many more buildings. Waves also hit Japan, 3,700 miles (6,000 km) away.

1979, Irian, Indonesia
Several earthquakes caused a series of tsunamis that destroyed many houses along the coast and killed about 100 people.

1979, Majuro, Marshall Islands
Two sets of tsunamis, a week apart, brought waves almost 23 feet (7 meters) high. The waves destroyed almost the entire city of Majuro, but no one was killed.

1979, San Juan Island, Colombia
A tsunami destroyed all of the houses on the island and 250 people drowned.

Glossary

aid help given as money, medicine, food, or other necessary items

broadcast program on radio or TV that gives information to many people

charity organization that gives out aid and makes people aware of disasters

continent seven large land masses on Earth—Asia, Africa, North America, South America, Europe, Australia, and Antarctica

coral reef hard substance built by colonies of tiny animals. After many years the coral builds up into a huge bank called a reef.

crest highest point

crop plant grown by people for food or other uses

debris broken pieces of buildings, trees, rocks, etc.

earthquake shaking of the ground caused by large movements inside Earth

electrocution when someone is injured by electricity

evacuate move away from danger until it is safe to return

first aid first medical help given to injured people

fjord narrow strip of sea between high cliffs

gravity force that attracts objects together and that holds us on the ground

headland strip of land that sticks out into the sea

landslide when a large piece of land suddenly slides down a slope

lava melted rock from inside the Earth that comes out of a volcano

livestock animals kept by people to eat or to sell

paramedic medical worker who travels to help people where an accident has happened

plate sheet of rock that forms part of the surface of Earth

polluted when air, soil, or water is poisoned or dirtied

powerline cable that carries electricity

reservoir large natural or man-made lake used to collect and store water

satellite object made by humans and put into space. Satellites do jobs such as sending out TV signals or taking photographs.

sea level normal level of the sea's surface and land that is at the same level

seafloor solid bottom of a sea or ocean

seismograph machine that records the force and direction of an earthquake

sewage waste matter from toilets and drains

simulate to accurately show how something would happen

trough wave at its lowest level

volcano hole in Earth's surface through which lava, hot gases, smoke, and ash escape

More Books to Read

Bonar, Samantha. *Tsunamis*. Mankato, Minn.: Capstone Press, 2001.

Flaherty, Michael. *Tidal Waves and Flooding*. Brooklyn, N.Y.: Millbrook Press, 1998.

Jennings, Terry. *Floods and Tidal Waves*. North Mankato, Minn.: Thameside Press, 1999.

Steele, Christy. *Tsunamis*. Austin, Tex.: Raintree Publishers, 2001.

Wade, Mary Dodson. *Tsunami: Monster Waves*. Berkeley Heights, N.J.: Enslow Publishers, 2002.

Index